SpringerBriefs in Fire

Series editor

James A. Milke, College Park, USA

More information about this series at http://www.springer.com/series/10476

Eddie Davis · Nick Kooiman
Kylash Viswanathan

Data Assessment for Electrical Surge Protective Devices

 Springer

Eddie Davis
Hughes Associates, Inc.
Vancouver, WA
USA

Kylash Viswanathan
Hughes Associates, Inc.
Vancouver, WA
USA

Nick Kooiman
Hughes Associates, Inc.
Vancouver, WA
USA

ISSN 2193-6595   ISSN 2193-6609   (electronic)
SpringerBriefs in Fire
ISBN 978-1-4939-2891-0   ISBN 978-1-4939-2892-7   (eBook)
DOI 10.1007/978-1-4939-2892-7

Library of Congress Control Number: 2015940729

Springer New York Heidelberg Dordrecht London

Printed on acid-free paper

Springer Science+Business Media LLC New York is part of Springer Science+Business Media
(www.springer.com)

Preface

Every year there are widespread anecdotal reports of homeowners' property damage to electrical and electronic equipment resulting from electrical surges. The revision cycle of the 2011 edition of NFPA 70, *National Electrical Code*® (NEC®) included several proposals (e.g., NEC 4-53 and NEC 4-127) to add new requirements for a Surge Protective Device for all dwelling units. These proposals were rejected by the respective Code Making Panel (i.e., CMP-4) due to a lack of reliable data to support such requirements.

Acknowledgments

The Research Foundation expresses gratitude to the report authors Eddie Davis, Nick Kooiman, and Kylash Viswanathan, all with Hughes Associates, Inc. Likewise, appreciation is expressed to the Project Technical Panelists and all others who contributed to this research effort for their ongoing guidance. Special thanks are expressed to the project sponsors, Eaton Corporation and the National Electrical Manufacturers Association, for providing the funding for this phase 1 project.

Technical Panel

Marty Ahrens, NFPA (MA)
Donny Cook, Shelby County & IAEI (AL)
Mark Earley, NFPA (MA)
Michael Johnston, NECA (MD)
Jack Jordan, State Farm (IL)
Andrew Trotta, CPSC (MD)

Sponsors

Eaton Corporation
National Electrical Manufacturers Association

Contents

Acronyms

μsec	Micro-second
EMI	Electromagnetic interference
FPRF	Fire Protection Research Foundation
GDT	Gas discharge tube
GPR	Ground potential rise
Hz	Hertz
III	Insurance Information Institute
kA	Thousands of amperes
khz	Kilo-hertz
L–G	Line-to-ground
L–L	Line-to-line
MCOV	Maximum continuous operating voltage
MOV	Metal oxide varistor
NEMA	National Electrical Manufacturers Association
NFPA	National Fire Protection Association
N–G	Neutral-to-ground
NIST	National Institute of Standards and Technology
NLDN	National Lightning Detection Network
pu	Per unit
SAD	Silicon avalanche diode
SCCR	Short circuit current rating
SPD	Surge protective device
TOV	Temporary overvoltage
TVSS	Transient voltage surge suppressor (no longer used—replaced by SPD)
UL	Underwriter's Laboratories
VPR	Voltage protection rating

List of Figures

List of Table

Chapter 1
Key Topics Explained

1.1 Surge Protection

Sources of Surges

A surge is a transient wave of voltage or current. The duration is not tightly specified but is usually less than a few milliseconds. The following are typical sources of surges:

- Lightning.
- Utility switching, including capacitor switching.
- Equipment switching and switching inductive loads within a facility.

Protection against surges is referred to as *surge protection*, and includes protection against both surge voltages and currents. The devices used to protect against surges are referred to as *surge protective devices*, or SPDs. A surge of duration longer than a few milliseconds is referred to as a swell or temporary overvoltage (TOV) and requires a different type of protection design; SPDs can fail if exposed to long-duration TOVs.

Surge Effects

Surges can cause equipment damage. Large surges damage equipment and other components in the electrical distribution system. Smaller surges can cumulatively damage equipment and can cause nuisance equipment tripping. Both surge voltage and current can be damaging. In the case of lightning strokes, the surge can be carried into a facility via all of the connected conductive paths.

There is a limit on how high a voltage can be transmitted into a facility or residence. Above a certain level, a high voltage will result in flashover in the insulation system of electrical equipment and conductors. A flashover can cause insulation damage, electric shock, and fire.

NEMA surveys of facility managers confirmed catastrophic failure or damage of electrical or electronic equipment due to a lightning event or voltage surge and

© Fire Protection Research Foundation 2015
E. Davis et al., *Data Assessment for Electrical Surge Protective Devices*,
SpringerBriefs in Fire, DOI 10.1007/978-1-4939-2892-7_1

premature failure of electrical or electronic equipment, including failure of life safety equipment.

The Insurance Information Institute report for 2013 identified 114,740 insurer-paid lightning claims for residential locations. The average lightning paid-claim amount was $5,869.

Residential Surge Protection
Residential surge protection has long been viewed as an important safety consideration and guidance has been issued by the IEEE and NIST to help homeowners protect their house and its contents. This protection has often been described as being similar to an insurance policy, partly because there is no NFPA code requirement for SPD installation in residences. Today's residences often contain electronic equipment throughout, including appliances, computers, security systems, life safety equipment, automation systems for Internet-enabled applications, and home entertainment systems.

1.2 Industry Standards

SPDs are routinely used in facilities that are potentially exposed to voltage or current surges from nearby lightning, utility switching, or other sources; and there are many manufacturers of SPDs who often offer guidance regarding SPD installation ratings and recommended applications. However, industry codes and standards provide limited guidance regarding selection, rating, and application.

One limitation with surge protection design is that there is no industry standard that describes what is an acceptable level of surge protection for standard facilities or residential locations. Although industry codes and standards are available that establish standardized surge criteria and assist with the application of specific surge protector types, these standards do not provide adequate design guidance that ensures a facility is properly protected against surges. There is no existing industry guidance for surge protection of residential facilities.

Refer to Chap. 4 for an overview of the available industry standards.

1.3 Data Acquisition Plan

Surge Data and Its Effects
Lightning strokes, either direct, nearby, or some distance away can cause voltage and current surges into a facility. Information is available regarding lightning strokes and their intensity. But, less information is available regarding the extent to which these surges are transmitted into commercial facilities, industrial facilities, or residences. Section 5.2 describes the difficulty associated with obtaining this data.

Available Data

Vaisala owns and operates the National Lightning Detection Network (NLDN) that provides accurate lightning data information across the USA. And, Vaisala can provide lightning location reports that provide individual cloud-to-ground lightning strikes and the intensity of strike at a specific location on the date of loss. This capability represents the largest and most complete source of lightning surge location and intensity.

Data for lightning surges that extend to the inside of facilities is not readily available. Published papers and IEEE C62.41.1 provide information regarding the expected surge levels within a facility or residence, but extensive data is not available.

Switching-related surge data, either internally or externally generated, is sparse. The added difficulty with this data is that these surges often do not cause immediate failure of electrical and electronic equipment; the damage occurs as a cumulative effect.

The largest documented source of surge effects is contained within the insurance claim documents for damage caused by surges. The Insurance Information Institute in collaboration with State Farm® produces annual reports of insurance claims associated with lightning-induced damage.

Data Acquisition Plan

There are challenges in obtaining usable data applicable to residential applications, such as:

- Confirming that equipment failures were a direct result of a surge event.
- Establishing any median and upper bounds to actual surge levels since this is not recorded inside facilities.
- Defining the protection improvement realized by applying SPDs.

Given the scarcity of real data relating to surges and the effects of surges, the approach described below is recommended.

The purpose of the recommended data acquisition approach is to produce real data regarding damage and injuries caused by surges. This information is intended to assist the NFPA 70 code-making committees with additional technical data to support a decision to require or not require SPDs for the variety of electrical applications proposed in past NFPA 70 update cycles (refer to Sect. 2.2).

The starting point for this project is to acquire the nationwide lightning stroke data for the continental USA for 2013 (or 2014 if the project starts in 2015). This information can tie back to insurance claim data and possibly provide surge current values for the locations of interest.

The Insurance Information Institute is proposed to manage the insurance industry claim data. Their involvement assures that the insurance industry claim reports can remain confidential while allowing access to additional data that might be contained in the claim reports.

The Insurance Information Institute already publishes annual summaries of the number of lightning-related insurance claims and the claim amount. Additional information of interest that might be available in the claim data includes:

- Date and location of surge event (to establish geographical correlations).
- Electronic equipment and appliances damaged.
- Life safety equipment damaged—smoke detectors, CO or CO_2 detectors, or other equipment.
- Fires caused by surge effects.
- Personal injuries associated with the surge event.
- Presence of or lack of installed SPDs.

Life safety equipment damage, fires caused by surge events, and personal injuries are of particular interest for code-making efforts.

Although the annual Insurance Information Institute survey has historically focused on residential claims, the survey for this project should include commercial and industrial claims also. NEMA assistance and direction with this effort will be helpful.

NEMA Low Voltage Surge Protective Devices Section (5-VS) participation is recommended for the following:

- Assisting with project scope, including commercial and industrial users.
- Reviewing the project checklist for the type of information to be obtained from the insurance industry.
- Reviewing failure data report summaries.
- Considering recommended SPD design principles, including the specification of surge protection in low-lightning flash density areas versus high-lightning flash density areas. Should NFPA elect to require SPDs in dwelling units or other applications, then minimum surge protection current limits should also be addressed, similar to the method provided in NFPA 780. As SPD surge current rating increases (and the degree of protection), the SPD cost also increases.

Chapter 2
Project Overview

2.1 Project Overview

The Fire Protection Research Foundation sponsored this project to address electrical surge protection for residential dwelling units. The goal of the project is to develop a data collection plan to assess loss related to electrical surges in homes, and address the potential impact electrical surge protection devices (SPDs) would have in mitigating these losses.

The project includes the following activities:

- Literature review—review of literature to include fundamental factors contributing to electrical surges, existing data associated with losses, case studies of SPD effectiveness, and overview of SPD designs.
- Preliminary data collection plan—develop a preliminary data collection plan that will address the identified data gaps. When implemented, the data collection plan should provide a comprehensive review of electrical surge related losses in homes in the United States and address the potential impact of electrical surge protection devices in mitigating these losses.

2.2 NFPA 70 Committee Report on Proposals—2013

Each update cycle for NFPA 70, *National Electrical Code*®, includes numerous proposals for changes throughout the document. In particular, the installation of SPDs has been proposed for virtually all low-voltage (600 V or less) electrical distribution equipment. Because of the breadth of these recommendations, the proposals and their reasons for rejection are summarized here. Although this Fire Protection Research Foundation report is focused on SPDs for residential dwelling units, the proposals for SPDs cover a much broader set of electrical distribution equipment.

© Fire Protection Research Foundation 2015
E. Davis et al., *Data Assessment for Electrical Surge Protective Devices*,
SpringerBriefs in Fire, DOI 10.1007/978-1-4939-2892-7_2

The *National Electrical Code® Committee Report on Proposals—2013 Annual Revision Cycle*[1] provides a summary of all proposals and their disposition in support of the 2014 edition of NFPA 70. With respect to the application of SPDs, the following proposals were submitted:

- Proposal 4-65 Log #3318 NEC-P04—New Article 225.41 Surge Protection. A Type 1 or Type 2 listed SPD shall be installed on all outside branch circuits and feeders and shall be located at the point where the outside branch circuits and feeders receive their supply.
- Proposal 4-143 Log #3319 NEC-P04—Article 230.67 Surge Protection. A Type 1 or Type 2 listed SPD shall be installed on all services.
- Proposal 4-143a Log #3504 NEC-P04—Article 230.67 Dwelling Unit Surge Protection.

  (A) Surge Protective Device. All dwelling units shall be provided with a surge protective device (SPD) installed in accordance with Article 285.
  (B) Location. The surge protective device shall be an integral part of the service disconnecting means or shall be located immediately adjacent thereto.
  (C) Type. The surge protective device shall be a Type 1 or Type 2 SPD.
  (D) Replacement. Where service equipment is upgraded, all of the requirements of this section shall apply.

- Proposal 5-244 Log #3320 NEC-P05—New Article 285.2 Required uses. A listed SPD shall be installed in or on the following equipment that is rated at 1000 V or less.

  (1) Switchboards and panelboards
  (2) Motor control centers
  (3) Industrial control panels
  (4) Control Panels for elevators, dumbwaiters, escalators, moving walks, platform and stairway chairlifts
  (5) Power distribution units supplying information technology equipment in information technology rooms
  (6) Solar photovoltaic (PV) combiner boxes, recombiner boxes, and inverters
  (7) Roof-top air conditioning and refrigerating equipment
  (8) Adjustable-speed drive systems
  (9) Burglar alarm panels
  (10) Fire alarm panels
  (11) Critical Operations Power Systems
  (12) Small Wind Electric Systems

- Proposal 9-117 Log #3321 NEC-P09—Article 408.6 Surge Protection. A listed SPD shall be installed in or on all switchboards and panelboards.

---

[1]The *National Electrical Code® Committee Report on Proposals—2013 Annual Revision Cycle*. The 2010 version provided similar recommendations.

- Proposal 11-14 Log #3322 NEC-P11—Article 409.70 Surge Protection. A listed SPD shall be installed in or on all industrial control panels.
- Proposal 11-42 Log #3323 NEC-P11—New Article 430.92 Surge Protection. A listed SPD shall be installed in or on all motor control centers.
- Proposal 11-55 Log #3324 NEC-P11—New Article 430.121 Surge Protection. A listed SPD shall be installed in or on all adjustable-speed drive systems.
- Proposal 11-84 Log #3325 NEC-P11—New Article 440.9 Surge Protection. A listed SPD shall be installed in or on all roof-top air-conditioning and refrigerating equipment.
- Proposal 12-49 Log #3326 NEC-P12—New Article 620.56 Surge Protection. A listed SPD shall be installed in or on control panels for elevators, dumbwaiters, escalators, moving walks, platform and stairway chairlifts.
- Proposal 12-140 Log #3327 NEC-P12—New Article 645.18 Surge Protection. A listed SPD shall be installed in or on all switchboards, panelboards, and power distribution units supplying information technology equipment in information technology rooms.
- Proposal 12-169 Log #3328 NEC-P12—New Article 670.6 Surge Protection. A listed SPD shall be installed in or on all industrial machinery.
- Proposal 4-254 Log #3329 NEC-P04—New Article 690.12 Surge Protection. A listed SPD shall be installed in or on all solar photovoltaic (PV) combiner boxes, recombiner boxes, and inverters.
- Proposal 13-98 Log #3330 NEC-P13—New Article 700.8 Surge Protection. A listed SPD shall be installed in or on all emergency systems switchboards and panelboards.
  Note: Although the *Committee Report on Proposals* lists the Final Action as Reject, the 2014 edition of NFPA 70 does include a new Article 700.8 that states:
  700.8 Surge Protection
  A listed SPD shall be installed in or on all emergency systems switchboards and panelboards.
- Proposal 4-405 Log #3331 NEC-P04—New Article 705.13 Surge Protection. A Type 1 listed SPD shall be installed at the point of connection of all interconnected electric power production sources.
- Proposal 3-131 Log #3332 NEC-P03—New Article 725.36 Surge Protection. A listed SPD shall be installed in or on all burglar alarm control panels.
- Proposal 3-179 Log #3333 NEC-P03—New Article 760.36 Surge Protection. A listed SPD shall be installed in or on all fire alarm control panels.

The NFPA 70 Panel rejected the above proposals on various bases, including:

- *Surge protection is permitted to be installed and should not be required, as surge probabilities vary by locality, and different types of electrical loads have differing surge protection requirements. Surge protection must also be periodically maintained or replaced. The user should make the decision to install this protection.*

- *While the use of SPD's is appropriate in many instances, it is not always needed in every installation. System designers should apply SPD's where needed. Equipment manufacturers frequently provide integrated surge protection when it is deemed appropriate. The substantiation provided does not warrant the imposition of this new requirement.*
- *Surge protective devices have proven to provide benefits for components and systems against the damages of voltage surges, but the substantiation for this proposal does not document that such protection would specifically benefit HVAC equipment installed on a roof. In addition this may not work with high resistance, impedance or ungrounded systems. The NFPA FPRF is working on a project in this area which may provide information in the future.*
- *CMP-13 acknowledges that surges may result in failures. However, the proposal does not state what type or level of protection should be required. Further substantiation through a formal research report that presents evidence of the type of SPD and the level of protection required would present the opportunity for the panel to reconsider the proposal.*

Miscellaneous changes were made to the 2014 edition of NFPA 70 Article 285, *Surge-Protective Devices (SPDs), 1000 V or Less*, but these changes do not affect the locations where surge protection has been required.

2.3 Report Content

This book provides information regarding:

- Surge phenomena and their sources.
- Surge protection methods.
- Surge protection strategies recommended by various sources.
- Industry standards and their recommendations.
- Available data associated with electrical surges and their impact.
- Recommended data collection in support of code-making efforts.

Chapter 3
Surge Protection Fundamentals

Abstract This chapter provides an overview of electrical surges and protection against the effects of these destructive surges.

3.1 Sources of Surges

A surge is a transient wave of voltage or current. The duration is not tightly specified but is usually less than a few milliseconds. The following are typical sources of surges:

- Lightning.
- Utility switching, including capacitor switching.
- Equipment switching and switching inductive loads within a facility.

The following summarizes the effects of these various surge sources (Table 3.1).

Protection against surges is referred to as *surge protection*, and includes protection against both surge voltages and currents. The devices used to protect against surges are referred to as *surge protective devices*, or SPDs. A surge of duration longer than a few milliseconds is referred to as a swell or temporary overvoltage (TOV) and requires a different type of protection design; SPDs can fail if exposed to long duration TOVs.

3.1.1 Lightning Surges

Lightning-induced surges into an electrical system are caused by lightning strokes to the ground, towers, or structures. A lightning stroke can produce peak discharge currents ranging from a few 1000–200,000 A, or higher. This lightning discharge current is developed within a few microseconds and typically discharges most of its

© Fire Protection Research Foundation 2015
E. Davis et al., *Data Assessment for Electrical Surge Protective Devices*,
SpringerBriefs in Fire, DOI 10.1007/978-1-4939-2892-7_3

Table 3.1 Sources of surges

Source of surge	Peak voltage magnitude	Frequency of occurrence	Comments
Lightning	<1,000 to >40,000 V with average of about 20,000 V	Weekly to rarely, depending on location	Magnitude depends on proximity of stroke to facility and coupling of stroke to facility electrical system. Voltages within a facility above 6,000 V are unlikely due to flashover
Utility capacitor and system switching	Up to 1,300 V on a 480 V system	Never to several times a day, depending on utility	Capacitors might or might not be installed nearby
Facility equipment switching	Up to 2,000 V on a 480 V system	Many times a day	Magnitude is small compared to lightning-induced transients, but switching can occur frequently

energy within a millisecond. The location where a lightning stroke will occur is not completely predictable; cloud-to-ground strokes have been recorded almost 20 miles from the base of the source cloud.

The frequency of lightning strokes varies with geographical location. Figure 3.1 shows the Vaisala lightning flash density map for the United States. Lightning strokes are a rare occurrence in Portland Oregon while they can be a routine event in Orlando Florida.

A single intense storm can produce thousands of lightning strokes. Schneider Electric Data Bulletin DB03A, *Surge Protection: Measured Lightning Stroke Data*, describes a July 2000 storm in Tampa Florida that recorded 33,863 lightning strokes during a 14 h period. Both positive and negative polarity strokes were detected,[1] with the following recorded surge currents:

Positive Lightning Stroke Surge Currents

- 95 %—less than 30 kA
- 98 %—less than 60 kA

Negative Lightning Stroke Currents

- 82 %—less than 30 kA
- 98 %—less than 60 kA

For this Tampa Florida storm, notice that the above results show that 2 % of the lightning strokes produced surge currents greater than 60 kA. A few lightning

---

[1] A lightning stroke is a lightning discharge between a thundercloud and the ground and commonly referred to as cloud-to-ground lightning. The most common type of lightning stroke is referred to as a negative lightning stroke and usually originates near the bottom of the cloud with a large concentration of negative charge in the cloud base. The term *negative lightning* means that there is a net transfer of negative charge from the cloud to the ground. Positive lightning strokes represent only about 5 % of the lightning strokes and tend to originate in the more positively charged top of the cloud.

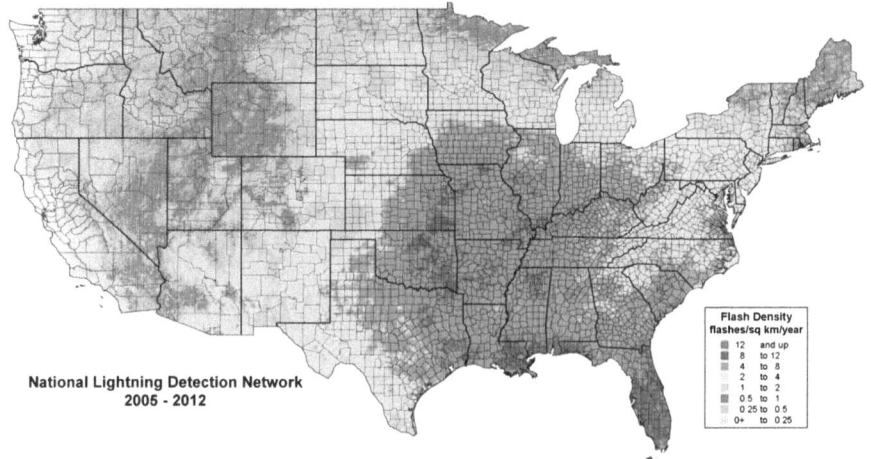

Fig. 3.1 Lightning flash density map. *Courtesy* Vaisala

strokes approached 200 kA. But, over 80 % of the lightning strokes produced surge currents less than 60 kA.

This data correlates reasonably well with a report from the IEEE Lightning and Insulator Subcommittee of the T&D Committee that showed a 50 % probability of less than about 20 kA, a 95 % probability of less than about 60 kA, and a 99 % probability of less than about 100 kA.[2]

A lightning-induced surge is a high magnitude impulsive transient of very short duration, typically measured in microseconds or milliseconds. But, during this short period, significant system damage can occur. Figure 3.2 shows an example.

Lightning-induced surges can be introduced into the electrical distribution system by any of the following methods, either alone or in combination[3]:

- Direct lightning strokes to the service entrance, either at low voltage lines or on the high voltage windings of service entrance transformers.
- Nearby strokes that induce voltages in distribution transformer secondary circuits.
- Strokes near the service entrance that induce surges onto the electrical system.
- Strokes to a building that induce surges in the system ground with respect to power supplies.
- Surges that cause surge protector operation, thereby placing a surge on the ground and neutral wire common to the low voltage system.

---

[2]Refer to Lightning and Insulator Subcommittee of the T&D Committee, *Parameters of Lightning Strokes: A Review*, IEEE Transactions on Power Delivery, Vol. 20, No. 1, January 2005, for the actual range of values.

[3]IEEE C62.41.1, *Guide On The Surge Environment In Low-Voltage (1000 V And Less) AC Power Circuits*, uses the terms "direct flash", "near flash", and "far flash" to distinguish between lightning strokes and how they induce a surge on a facility.

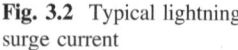

Fig. 3.2 Typical lightning
surge current

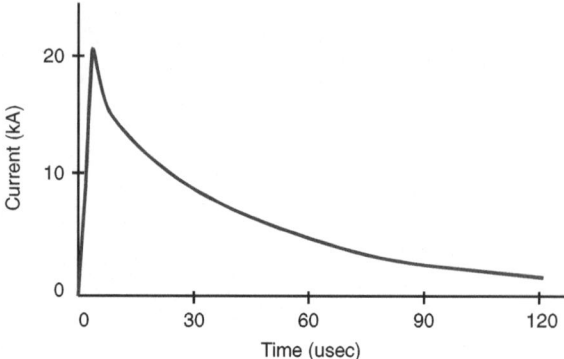

The greatest number of lightning-caused surges that will be seen originate on the high voltage side of the distribution transformer. Far less often, the surges will be caused by a local stroke impinging on the facility, service entrance transformer, or nearby equipment. Most surges, regardless of whether they originate on the primary or secondary side of the transformer are not from a direct stroke; usually, the surge is caused by a stroke to the pole, tower, ground wire, or nearby object with the surge electromagnetically coupled into the distribution or service conductors. Once into the electrical system wiring, surges on the high side of the transformer are coupled into the secondary and transmitted throughout a facility.

3.1.2 Utility Switching

Utility switching is a broad term that applies to how utility configurations are occasionally changed. Each switching operation can produce a transient that can momentarily exceed equipment voltage ratings. Although the transients are not as large in magnitude as a nearby lightning stroke, switching transients can cause cumulative damage to electrical equipment. And, if switching results in a temporary overvoltage (TOV), it can also cause SPD failure.

Capacitor switching is a special case of utility switching. Capacitors might also be switched periodically by large industrial power customers. Capacitor switching can be a common every-day event, occurring several times each day in some locations, as a utility adjusts system voltage and compensates for inductive loads.

Capacitor switching causes a surge voltage by the following process. The voltage across a capacitor is zero before it is switched into the circuit. As a capacitor is switched, there is a momentary short circuit across the capacitor as the system voltage is applied to the zero voltage of the capacitor. At the capacitor location, the bus voltage momentarily experiences a step change to zero volts. After the initial step change, the voltage recovers and then overshoots as the system eventually return to its steady state value. Thereafter, the system oscillates until damping returns the voltage to its steady-state value. During the initial oscillation period, the peak

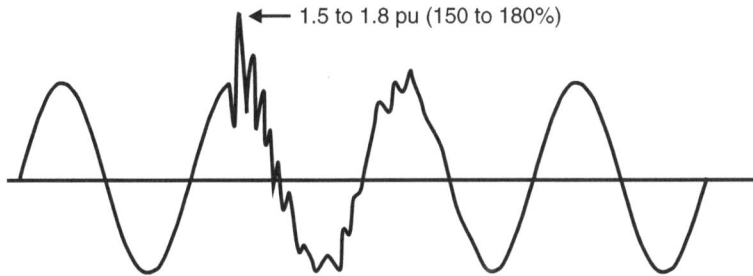

Fig. 3.3 Voltage waveform for capacitor switching transient

transient voltage can approach 200 % of the normal peak system voltage (common peak surge voltages can range from 150 to 180 % of normal). Another factor contributing to the transient is the inrush current as the capacitor energizes; this inrush current can have a resonant frequency anywhere from 300 to 1,000 Hz depending on the installed inductance and capacitance, which adds to the system oscillations. Figure 3.3 shows an example of a capacitor switching transient.

Capacitors inside a facility can resonate with the switching-induced oscillations, thereby magnifying the peak voltage and extending the period until the voltage returns to normal. Magnification of the switching transient can occur if the utility switched capacitor bank is much larger than the facility capacitor bank and there is little resistive load (mostly motor load) to provide a damping mechanism.

3.1.3 Facility Internal Switching

Switched equipment in a facility electrical system or residence results in the inductive release of energy that creates a momentary voltage surge. Even minor switching, such as deenergizing lighting loads, can cause a significant inductive surge in the system. This type of switching accounts for the overwhelming majority of switching transients. However, the magnitude of this type of surge is much smaller than for lightning-induced surges.

3.2 Surge Effects

Surges can cause equipment damage.[4] Large surges damage equipment and other components in the electrical distribution system. Smaller surges can cumulatively damage equipment and can cause nuisance equipment tripping. Both surge voltage

---

[4]Refer to IEEE 1100, *Powering and Grounding Electronic Equipment*, for additional information regarding the effects of surges.

and current can be damaging. In the case of lightning strokes, the surge can be carried into a facility via all of the connected conductive paths. The following figures show examples of damage caused by surges (Figs. 3.4, 3.5, 3.6 and 3.7).

Electronic equipment is susceptible to surge transients. Computers and internet-enabled devices are not only at risk in the power supply but can also be damaged by surges that propagate into the equipment via the communications link.

There is a limit on how high of a voltage can be transmitted into a facility or residence. Above a certain level, a high voltage will result in flashover in the insulation system of electrical equipment and conductors. A flashover can cause insulation damage, electric shock, and fire.

Fig. 3.4 Circuit breaker failure caused by surge voltage

Fig. 3.5 Copper busbar melted by surge current

Fig. 3.6 Circuit board
damage caused by surge
voltage

3.2.1 NEMA Surveys

The NEMA Low Voltage Surge Protective Devices Section (5-VS) sponsored
surveys of surge damage in 2013 and 2014.[5] The surveys were targeted towards
facility managers and attempted to accomplish the following:

- Determine if SPDs are installed.
- Obtain failure data for electrical or electronic equipment due to a lightning event
  or voltage surge.
- Determine the frequency of damage incidents.
- Address the type of equipment damaged.
- Summarize the cost of damage.

---

[5]NEMA 2013 *U.S. Surge Protection Damage Survey* and NEMA *Surge Damage Survey Results—
Wave 2*. Refer to http://www.nemasurge.org for reports.

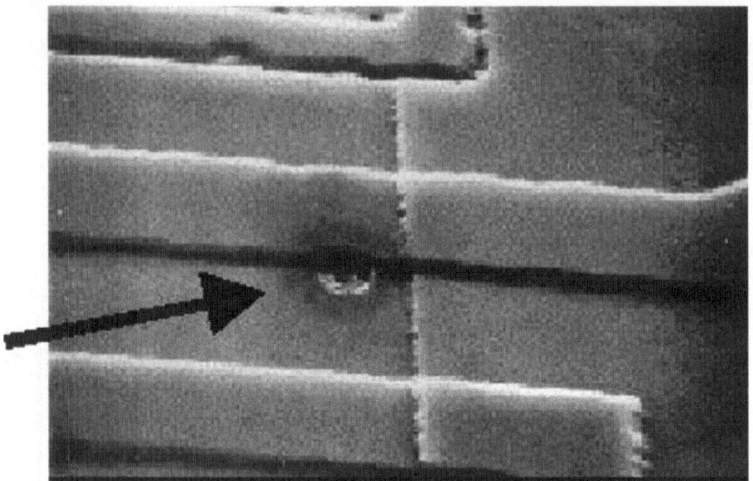

Fig. 3.7 Micro circuit damage caused by surge voltage

The following summarizes the 2014 survey results:

- 100 respondents completed the survey.
- A plurality (48 %) of respondents noted that their facility had experienced unexplained process interruptions. Catastrophic failure or damage of electrical or electronic equipment due to a lightning event or voltage surge and premature failure of electrical or electronic equipment were both frequently reported (41 %) events. More than a third (38 %) noted the occurrence of lockup of computer or industrial process systems.
- For most respondents (61 %), it cost less than $10,000 to repair the damage resulting from voltage surges, but a sizable number (16 %) reported damage costing in excess of $50,000 to fix.
- Nearly 95 % of those who reported having experienced a surge event resulting in equipment damage indicated that they subsequently purchased surge protection. Virtually all of those who did so, purchased immediately or within three months of the event.
- Over 65 % reported downtimes associated with voltage surges of 6 h or more.
- Respondents reported damage or loss of function of the following types of life safety equipment because of voltage surges:

  - Smoke detector (34.7 %)
  - CO_2 detector (18.7 %).
  - Fire alarm system (41.3 %).
  - Security system (49.3 %).
  - Ground fault circuit interrupters (22.7 %).
  - Emergency lighting (32.0 %)
  - Emergency generators or backup power (33.3 %).

- Fire pumps (12.0 %).
- Elevators or escalators (24.0 %).
- Safety interlocking systems on machines (26.7 %)

Of the respondents, only 14.7 % stated that no life safety equipment was damaged or lost function.

- When asked if anyone had been injured, either directly or indirectly, as a result of a voltage surge, 10.7 % replied yes.

The NEMA survey is significant in that it shows the effect of surges on life safety equipment and the potential impact to personnel in a facility.

3.2.2 Insurance Information Institute Surveys

The Insurance Information Institute[6] provides periodic reports of homeowner insurance claims associated with lightning-induced damage. Their report for 2013, produced in collaboration with State Farm®, included the following:

- There were 114,740 insurer-paid lightning claims in 2013, down 24 % from 2012.
- The average lightning paid-claim amount was also down in 2013, slipping by 8.3 % to $5,869 from $6,400 in 2012.
- The decline in lightning damage last year is consistent with data from the National Weather Service, which recorded 137 days in 2013 with lightning causing property damage, while 160 such days were recorded in 2012—a 14 percent decrease.
- Despite the drop in the number of paid claims in 2013, the average cost per claim rose nearly 122 % from 2004 to 2013. The average cost per claim has generally continued to rise, in part because of the huge increase in the number and value of consumer electronics in homes.

3.3 Surge Protective Devices (SPDs)

3.3.1 Typical Configuration

Most SPDs in use for the applications covered by this book use metal oxide varistors (MOVs) to accomplish surge suppression in the electrical power system. MOVs exhibit nonlinear resistance characteristics as a function of voltage. Within the MOV voltage rating, the resistance usually exceeds 10,000,000 Ω, but the resistance drops to less than 0.1 Ω when the MOV is exposed to an overvoltage,

---

[6]Their reports are accessible at http://www.iii.org/.

such as a transient voltage spike due to a nearby lightning stroke. It is this characteristic that makes MOVs an effective protection element.

The MOV is essentially a matrix of zinc oxide grain boundaries that have a nonlinear resistance characteristic. The series combination of the boundaries defines the MOV voltage rating, the parallel combination defines the total current that can be passed, and the bulk volume determines how much energy that it can absorb. When an MOV is energized with an AC voltage, resistive and reactive current flows through the highly capacitive disc.

Most SPDs are connected in parallel with the circuit and operate when a transient voltage exceeds the voltage protection rating. Parallel surge protectors have little interaction with the circuit under normal conditions.

A different technology is commonly used for communications lines, referred to as a gas discharge tube (GDT), which is a spark gap type of surge suppression device. When subjected to a surge voltage, the gas discharge tube sparks over, thereby causing an arc to ground. The hermetically sealed tubes used today can have a precise and repeatable turn-on voltage. Gas discharge tubes consist of a spark gap in series with a resistance or varistance to limit the discharge current to safe levels.

3.3.2 SPD Classification

UL 1449 classifies SPDs by type depending, in part, on their location in the system and their level of internal protection:

- Type 1—Permanently connected SPDs intended for installation between the secondary of the service transformer and the line side of the service equipment overcurrent device, as well as the load side, including watt-hour meter socket enclosures and intended to be installed without an external overcurrent protective device. They must have overcurrent protective devices either installed internally on the SPD or included with it. While these are primarily intended for installation before the main service disconnect, Type 1 SPDs can be installed in Type 2 and Type 4 locations such as distribution panels, end-use equipment. Residential installations are often Type 1, installed near the incoming meter.
- Type 2—Permanently connected SPDs intended for installation on the load side of the service equipment overcurrent device; including SPDs located at the branch panel. While some will have internal overcurrent protective components, Type 2 SPDs can rely on the service entrance overcurrent disconnect device for over current protection. These SPDs can be installed in service equipment, distribution panels, and end-use equipment.
- Type 3—Point of utilization SPDs, installed at a minimum conductor length of 10 m (30 feet) from the electrical service panel to the point of utilization, for example cord connected, direct plug-in, receptacle type and SPDs installed at the utilization equipment being protected.

- Type 4—Component SPDs, including discrete components as well as component assemblies.

Permanently installed self-contained SPDs are usually Type 1 or Type 2.

3.3.3 SPD Ratings

SPDs are tested and rated in accordance with UL 1449. The following ratings are normally provided for each model and size of SPD:

- Nominal voltage and frequency.
- Maximum continuous overvoltage (MCOV)—defines the voltage at which the SPD will start conducting to ground. Continuous operation above the MCOV will lead to SPD failure.
- Voltage protection rating (VPR)—a UL 1449 rating of the limiting voltage measured during the transient-voltage surge suppression test using the combination wave generator at a setting of 6 kV, 3 kA. A lower VPR is better.
- Surge current rating—the maximum surge current that an SPD is rated to carry without excessive overheating and consequent premature breakdown or combustion risk. The surge current rating is expressed in thousands of amps (kA) and is an indicator of how many MOVs are installed in parallel inside the device. SPDs are readily available rated for as low as ≤20 kA up to ≥600 kA. SPD price tends to increase as surge current rating increases.
- Protection modes—line-to-line, line-to-ground, line-to-neutral, neutral-to-ground.
- Short circuit current rating (SCCR).
- Surge life—expected number of surges that the SPD can withstand.

Other important attributes include monitoring and design for the environment at the installation location.

3.4 Residential Surge Protection

Residential surge protection has long been viewed as an important safety consideration and guidance has been issued in the past to help homeowners protect their house and its contents.[7] This protection has often been described as being similar to an insurance policy, partly because there is not an NFPA code requirement for SPD

---

[7]Key documents include *How to Protect Your House and Its Contents from Lightning, IEEE Guide for Surge Protection of Equipment Connected to AC Power and Communication Circuits*, by Richard L. Cohen and others, ISBN 0-7381-4634-X, 2005 and NIST Special Publication 960–6, *Surges Happen! How to Protect the Appliances in Your Home*, 2001. Some insurance companies also provide guidance on their websites.

installation in residences. Today's residences often contain electronic equipment throughout, including appliances, computers, security systems, life safety equipment, automation systems for internet-enabled applications, and home entertainment systems.

3.4.1 Design

SPDs used in residential applications are typically designed for 240/120 V with the electrical power neutral bonded to ground at the service entrance. A permanently-installed SPD can be installed at the incoming meter (Type 1) or at the service entrance (Type 2). Type 3 SPDs can still be installed at the point of use for electronic equipment also.

The IEEE document, *How to Protect Your House and Its Contents from Lightning, IEEE Guide for Surge Protection of Equipment Connected to AC Power and Communication Circuits*, provides an excellent overview of the design and installation considerations for SPDs. A permanently-installed SPD should be installed by a qualified electrician and should consider quality of the grounding system, lead length for connections, overcurrent protection, and disconnect capability. Installation in accordance with NFPA 70 is a requirement.

3.4.2 General Cost

Prices vary widely for SPDs. An SPD intended for residential use (240/120 V) and rated for 50 kA surge current can cost as little as $125 and as much as $500. Integrated protection to protect the incoming power lines as well as the phone/internet communication lines can cost an additional $100. A reasonable level of protection can typically be realized for about $500.

One consideration is how high of a surge current rating to specify. Cost tends to increase as the surge current rating increases because of the additional MOV modules that are required. The cost can be considerably higher for three-phase circuits, partly because there are more protection modes to consider compared to a single-phase application and partly because the surge current rating might be higher. For residential applications, a surge current rating above 30 kA likely is adequate for 80–90 % of lightning strokes.[8] A surge current rating above 60 kA likely is adequate for virtually all lightning strokes. In lightning-prone areas (refer to Fig. 3.1), a higher surge current rating can also provide a longer SPD life if it is exposed to repeated surges.

---

[8]Lightning strokes produce the largest surges.

Chapter 4
Industry Standards

Abstract This chapter provides an overview of industry codes and standards that apply to SPDs. SPDs are routinely used in facilities that are potentially exposed to voltage or current surges from nearby lightning, utility switching, or other sources; and there are many manufacturers of SPDs and these manufacturers often offer guidance regarding SPD installation ratings and recommended applications. However, industry codes and standards provide limited guidance regarding selection, rating, and application. One limitation with surge protection design is that there is no industry standard that describes what is an acceptable level of surge protection for standard facilities or residential locations. Although industry codes and standards are available that establish standardized surge criteria and assist with the application of specific surge protector types, these standards do not provide adequate design guidance that ensures a facility is properly protected against surges. There is no existing industry guidance for surge protection of residential facilities.

4.1 NFPA Codes and Standards

4.1.1 NFPA 70

NFPA 70 distinguishes between surge arresters for applications over 1,000 V (Article 280) and SPDs for applications 1,000 V or less (Article 285). Each Article provides installation requirements.

NFPA 70[1] requires SPDs for the following applications:

- Article 501.35, Surge Protection—required Class I Division 1 and 2 locations.
- Article 694, Wind Electric Systems. Article 694.7(D) requires, "*A surge protective device shall be installed between a small wind electric system and any*

---

[1] The *National Electrical Code*® *Handbook* provides additional discussion of surge protection requirements.

© Fire Protection Research Foundation 2015

E. Davis et al., *Data Assessment for Electrical Surge Protective Devices*, SpringerBriefs in Fire, DOI 10.1007/978-1-4939-2892-7_4

loads served by the premises electrical system. The surge protective device shall be permitted to be a Type 3 SPD on a dedicated branch circuit serving a small wind electric system or a Type 2 SPD located anywhere on the load side of the service disconnect."

• Article 700, Emergency Systems. New Article 700.8, Surge Protection, was added in 2014 and requires, *"A listed SPD shall be installed in or on all emergency systems switchboards and panelboards."*
• Article 708, Critical Operations Power Systems (COPS). Article 708.20(D) requires, *"Surge protection devices shall be provided at all facility distribution voltage levels".*
• If surge protection is provided, Article 646, Modular Data Centers, requires that SPDS are listed, labeled, and installed in accordance with Article 285.

Article 285 provides requirements regarding the installation of SPDs, but it provides limited guidance for when a SPD is required or recommended ratings. The NFPA 70 Handbook also avoids discussion regarding the application of SPDs. In other words, NFPA 70 provides guidance regarding SPD installation, but provides no information regarding SPD selection and rating.

4.1.2 NFPA 780

NFPA 780,[2] *Standard for the Installation of Lightning Protection Systems*, is more specific regarding the application of SPDs for lightning protection systems. This standard provides detailed requirements for the application of SPDs in support of a lightning protection system, including SPD rating information. Key requirements include:

• SPDs shall be installed at all power service entrances.
• The SPD shall protect against surges produced by a 1.2/50 μs and 8/20 μs combination waveform generator.
• SPDs at the service entrance shall have a nominal discharge current (I_n) rating of at least 20 kA 8/20 μs per phase.
• Signal, data, and communications SPDs shall have a maximum discharge current (I_{max}) rating of at least 10 kA 8/20 μs when installed at the entrance.
• The published voltage protection rating (VPR) for each mode of protection shall be selected to be no greater than those given in Table 4.20.4 for the different power distribution systems to which they can be connected. The maximum allowed VPR per mode of protection varies from 600 to 1,800 V, depending on the service voltage and connection type.

[2]NFPA 780, *Standard for the Installation of Lightning Protection Systems*, 2014 Edition.

- The maximum continuous operating voltage (MCOV) of the SPD shall be selected to ensure that it is greater than the upper tolerance of the utility power system to which it is connected.
- SPDs at grounded service entrances shall be wired in a line-to-ground (L–G) or line-to-neutral (L–N) configuration. Additional modes, line-to-line (L–L), or neutral-to-ground
- (N–G) shall be permitted at the service entrance. For services without a neutral, SPD elements shall be connected line-to-ground (L–G). Additional line-to-line (L–L) connections shall also be permitted.
- Installation of surge suppression hardware shall conform to the requirements of NFPA 70, *National Electrical Code*. SPDs shall be located and installed so as to minimize lead length. Interconnecting leads shall be routed so as to avoid sharp bends or kinks.

Although NFPA 780 only applies to lightning protection systems, it provides clear SPD design, rating, and installation guidance for these systems.

4.2   IEEE Standards

IEEE has historically taken the lead with respect to characterizing the surge environment. The following sections discuss key IEEE documents that apply to SPDs.

4.2.1   IEEE C62.41.1

IEEE C62.41.1, *Guide On The Surge Environment In Low-Voltage (1000 V And Less) AC Power Circuits*, provides comprehensive information about surges and the environment in which they occur. This guide form the basis for IEEE surge testing criteria and is recommended for any review of surge characteristics. IEEE C62.41.1 is also valuable as a source of recorded data of surge events. Temporary overvoltages are also discussed, including their potential impact on SPDs.

4.2.2   IEEE C62.41.2

IEEE C62.41.2, *Recommended Practice On Characterization Of Surges In Low-Voltage (1000 V And Less) AC Power Circuits*, presents recommendations for selecting surge waveforms, and the amplitudes of surge voltages and currents used to evaluate equipment immunity and performance of SPDs. Figures 4.1, 4.2 and 4.3 show the surges recommended by IEEE C62.41.2 for consideration.

The second type of IEEE C62.41.2 surge voltage is called a 100 kHz ring wave with a waveform below.

Fig. 4.1 Combination
wave—1.2 × 50 μs, open
circuit voltage

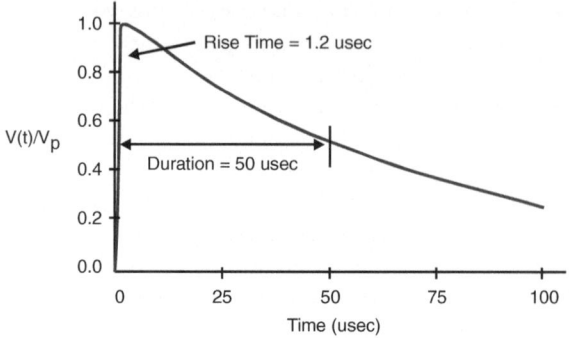

Fig. 4.2 Combination
wave—8 × 20 μs, short circuit
current

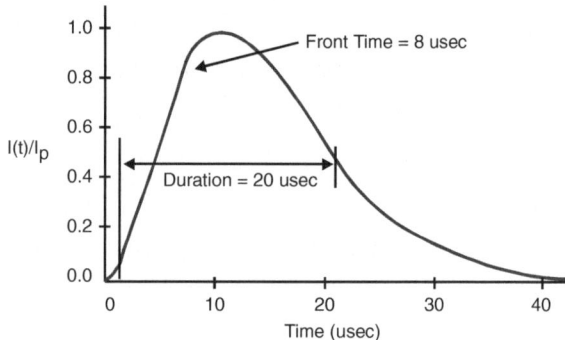

Fig. 4.3 100 kHz ring
wave—open circuit voltage

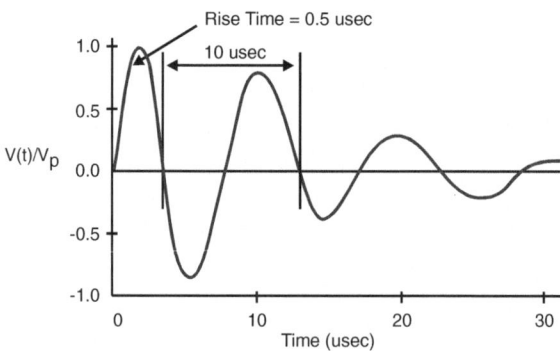

The combination and ring waves are intentionally generic in shape, in that peak magnitudes are not provided. Voltage and current values are assigned according to distance into the distribution system.

4.2.3 IEEE C62.45

IEEE C62.45, *Recommended Practice On Surge Testing For Equipment Connected To Low-Voltage (1000 V And Less) AC Power Circuits*, describes surge testing procedures using simplified waveform representations (described in IEEE C62.41.2) to obtain reliable measurements and enhance operator safety.

4.2.4 IEEE 1100

IEEE 1100, *Powering and Grounding Electronic Equipment*, provides guidance regarding SPDs. Unfortunately, the information provided in IEEE 1100 is dated and does not reflect the current SPD products that are available; much of the information provided is over 15 years old. But, IEEE 1100 provides a good discussion of surge effects and protecting against surges.

4.2.5 IEEE 1692

IEEE 1692, *IEEE Guide for the Protection of Communication Installations from Lightning Effects*, provides design guidelines to help prevent lightning damage to communications equipment within structures.

4.3 UL Documents

4.3.1 UL 1449

In the USA, SPDs are manufactured and specified in accordance with UL 1449, Third Edition, *Surge Protective Devices*, which was issued on September 29, 2006 with an effective date of September 29, 2009. This revision to UL 1449 changed how surge protective devices (SPDs) are named, tested, and rated. UL 1449 listing is specifically required by NFPA 780 and SPD listing (presumably to UL 1449) is required by NFPA 70. UL 1449 defines the performance requirements for an SPD; however, it does not address the engineering application of SPDs. UL also addresses additional related product performance criteria in UL 1283, *Electromagnetic Interference Filters*, and the UL 497 series, *Protectors for Fire Alarm Signaling Circuits*.

UL 1449, Third Edition, improved the harmonization of methods with IEC 61643 series, *Low Voltage Surge Protective Devices*, but there still remain some differences in approach between the UL and IEC test methods.

UL 1449, Third Edition, changed testing and rating requirements such that an SPD listed to UL 1449, Second Edition, cannot be compared to an SPD listed to UL 1449, Third Edition; the differences are too significant. Some of the key changes include:

- New performance tests use more surge current, resulting in higher let-through voltages. The older tests were performed at 500 A and 6,000 V. The new tests are performed at 3,000 A and 6,000 V.
- Test results for the new performance tests in the Third Edition are higher than the equivalent tests in the Second Edition, which has resulted in manufacturers changing their product literature. With a surge current of 6 times the Second Edition level, the Third Edition results for let-through voltage must be higher (the let-through voltage or clamping voltage was referred to as suppressed voltage rating in the Second Edition and is referred to as voltage protection rating in the Third Edition).
- Terminology has changed.
- UL 1449 is now ANSI-approved.

4.3.2 UL 497

The UL 497 series, *Protectors for Fire Alarm Signaling Circuits*, provides performance standards and testing procedures for enclosures, corrosion protection, field wiring connections, and components of SPDs, as well as product labeling and installation instructions.

4.3.3 UL 1283

UL 1283, *Electromagnetic Interference Filters*, provides requirements for electromagnetic interference (EMI) filters. It addresses filters installed on, or connected to, 600 V or lower voltage circuits and 50–60 Hz frequency. These filters are used to attenuate unwanted radio frequency (RF) signals, such as noise or interference generated from electromagnetic sources. They consist of capacitors and inductors used alone or in combination with each other and may be provided with resistors.

Chapter 5
Data Acquisition Plan

Abstract This chapter provides an overview of the recommended data acquisition plan for SPDs.

5.1 Type of Desired Data

In order to address fully the potential application of SPDs as a code requirement, the following types of data would be especially helpful:

Installations With SPDs Installed

- Characterization of surge events that were successfully diverted without damage to electronic equipment, electrical equipment, or the structure.
- Characterization of surge events that occurred with subsequent damage to electronic equipment, electrical equipment, or the structure.
- Characterization of surge events that resulted in damage or loss of function to life safety equipment.

Installations Without SPDs Installed

- Characterization of surge events that did not cause damage to electronic equipment, electrical equipment, or the structure.
- Characterization of surge events that occurred with subsequent damage to electronic equipment, electrical equipment, or the structure.
- Characterization of surge events that resulted in damage or loss of function to life safety equipment.

Surge Characterization

- Real data recording of lightning-induced surges.
- Real data recording of switching-induced surges, either internally generated (appliances or motors turning on or off) or externally generated (such as capacitor switching).

© Fire Protection Research Foundation 2015

E. Davis et al., *Data Assessment for Electrical Surge Protective Devices*,
SpringerBriefs in Fire, DOI 10.1007/978-1-4939-2892-7_5

The problem lies in acquiring the above data, which is the goal of this project. The above information does not really exist, except in a few limited scope studies and in insurance claim documents. Refer to the following sections for more information.

5.2 Data to Characterize the Nature of Surges

Lightning strokes, either direct, nearby, or some distance away can cause voltage and current surges into a facility. Information is available regarding lightning strokes and their intensity. But, less information is available regarding the extent to which these surges are transmitted into commercial facilities, industrial facilities, or residences. The IEEE paper, *A Field Study of Lightning Surges Propagating Into Residences,*[1] provides an outstanding view into the effort needed to acquire even limited amounts of real-world data. This paper provides the following insights:

- When a home appliance malfunctions due to lightning, the relationship between the lightning stroke and the damage is often unclear. The purpose of their study was to complete experimental investigations on lightning surges that flow into residences. SPDs were not installed in these residences.
- Lightning surge waveform detectors were installed in 49 residences and monitored for 4 years (2003–2006). During the 4 year observation period, lightning surge waveforms were obtained for a total of 18 lightning stroke events.
- Damage occurred to appliances in 4 of the 18 events. The most severe damage occurred when lightning appeared to have hit an antenna. In this case, currents of 1 kA or greater were recorded at all the measurement points, and many appliances were damaged.
- The home appliances, typically having built-in lightning protective devices with a peak current of 1 kA or higher, broke down at a current peak value of approximately 1 kA or higher, according to the observations.
- The analysis of observation data found that in some cases a ground potential rise causes a lightning surge to flow from the ground of another residence or the ground of a distribution system into the distribution system and, in turn, to flow into another residence.

Notice that the above effort took 4 years of monitoring at 49 residences to produce recordings of 18 surge events, of which four were severe enough to cause damage to appliances. This illustrates the difficulty of acquiring actual surge data.

---

[1]Teru Miyazaki et al. *A Field Study of Lightning Surges Propagating Into Residences*, IEEE Transactions on Electromagnetic Compatibility, Vol. 52, No. 4, November 2010.

5.3 Who Has Data on Surges, Surge Effects, and SPDs

5.3.1 Surge Data—Lightning Surges

Vaisala[2] owns and operates the National Lightning Detection Network (NLDN) that provides accurate lightning data information across the USA. And, Vaisala can provide lightning location reports that provide individual cloud-to-ground lightning strikes and the intensity of strike at a specific location on the date of loss. This capability represents the largest and most complete source of lightning surge location and intensity.

Data for lightning surges that extend to the inside of facilities is not readily available. Published papers and IEEE C62.41.1 provide information regarding the expected surge levels within a facility or residence, but extensive data is not available.

5.3.2 Surge Data—Switching Surges

Switching-related surge data, either internally or externally generated, is sparse. The added difficulty with this data is that these surges often do not cause immediate failure of electrical and electronic equipment; the damage occurs as a cumulative effect.

5.3.3 Surge Effects—Manufacturers

Although manufacturers produce SPDs and do a great job of educating consumers regarding their products, very little failure data associated with surges appears to be available from them. NEMA maintains the Surge Protection Institute[3] and they have completed surveys in 2013 and 2014 regarding failures of electrical and electronic equipment caused by surges. Although the sample size is relatively small, the survey results are helpful with respect to historical failures of life safety equipment. Refer to Sect. 3.2.1 for more information.

5.3.4 Surge Effects—Consulting Firms

Many engineering consulting firms assist with evaluations of surge damage in support of insurance claims. However, this data is not compiled in a readily usable manner nor is the data accessible in many cases. Surge data is not typically available.

---

[2]http://www.vaisala.com/en/services/dataservicesandsolutions/lightningdata/Pages/default.aspx.

[3]*Refer to* http://www.nemasurge.org.

5.3.5 Surge Effects—Insurance Claims

The largest documented source of surge effects is contained within the insurance claim documents for damage caused by surges. The Insurance Information Institute in collaboration with State Farm® produces annual reports of insurance claims associated with lightning-induced damage (refer to Sect. 3.2.2 for more information). The information contained in these claim reports likely provides additional detail regarding surge effects and the types of damage caused.

5.4 Data Acquisition Plan

There are challenges in obtaining usable data applicable to residential applications, such as:

- Confirming that equipment failures were a direct result of a surge event.
- Establishing any median and upper bounds to actual surge levels since this is not recorded inside facilities.
- Defining the protection improvement realized by applying SPDs.

Given the scarcity of real data relating to surges and the effects of surges, the approach described below is recommended.

5.4.1 Purpose of Data to Be Obtained

The purpose of the recommended data acquisition approach is to produce real data regarding damage and injuries caused by surges. This information is intended to assist the NFPA 70 code-making committees with additional technical data to support a decision to require or not require SPDs for the variety of electrical applications proposed in past NFPA 70 update cycles (refer to Sect. 2.2).

5.4.2 Lightning Stroke Data

The starting point for this project is to acquire the nationwide lightning stroke data for the continental USA for 2013 (or 2014 if the project starts in 2015). This information can tie back to insurance claim data and possibly provide surge current values for the locations of interest.

5.4.3 Insurance Information Institute Claim Data

The Insurance Information Institute is proposed to manage the insurance industry claim data. Their involvement assures that the insurance industry claim reports can remain confidential while allowing access to additional data that might be contained in the claim reports.

The Insurance Information Institute already publishes annual summaries of the number of lightning-related insurance claims and the claim amount. Additional information of interest that might be available in the claim data includes:

- Date and location of surge event (to establish geographical correlations).
- Electronic equipment and appliances damaged.
- Life safety equipment damaged—smoke detectors, CO or CO_2 detectors, or other equipment.
- Fires caused by surge effects.
- Personal injuries associated with the surge event.
- Presence of or lack of installed SPDs.

Life safety equipment damage, fires caused by surge events, and personal injuries are of particular interest for code-making efforts.

Although the annual Insurance Information Institute survey has historically focused on residential claims, the survey for this project should include commercial and industrial claims also. NEMA assistance and direction with this effort will be helpful.

5.4.4 NEMA Participation

NEMA Low Voltage Surge Protective Devices Section (5-VS) participation is recommended for the following:

- Assisting with project scope, including commercial and industrial users.
- Reviewing the project checklist for the type of information to be obtained from the insurance industry.
- Reviewing failure data report summaries.
- Considering recommended SPD design principles, including the specification of surge protection in low-lightning flash density areas versus high-lightning flash density areas. Should NFPA elect to require SPDs in dwelling units or other applications, then minimum surge protection current limits should also be addressed, similar to the method provided in NFPA 780. As SPD surge current rating increases (and the degree of protection), the SPD cost also increases.

5.4.5 Why not Another Test Program?

The IEEE paper, *A Field Study of Lightning Surges Propagating Into Residences*, illustrates the difficulty with obtaining real data during surge events. Although this study produced very useful results, it took a 4-year period at 49 homes to obtain data for 18 surge events, of which four surge events caused damage to appliances and electronic equipment. This is considered a typical outcome to be expected. A test program sponsored by the Fire Protection Research Foundation is not recommended.

References

A.1 Industry Standards

IEEE 1100, *Powering and Grounding Electronic Equipment*.
IEEE 1692, *IEEE Guide for the Protection of Communication Installations from Lightning Effects*.
IEEE C62.41.1, *Guide On The Surge Environment In Low-Voltage (1000 V And Less) AC Power Circuits*.
IEEE C62.41.2, *Recommended Practice On Characterization Of Surges In Low-Voltage (1000 V And Less) AC Power Circuits*.
IEEE C62.45, *Recommended Practice On Surge Testing For Equipment Connected To Low-Voltage (1000 V And Less) AC Power Circuits*.
IEEE C62.50, *IEEE Standard for Performance Criteria and Test Methods for Plug-in (Portable) Multiservice (Multiport) Surge-Protective Devices for Equipment Connected to a 120 V/240 V Single Phase Power Service and Metallic Conductive Communication Line(s)*.
NFPA 70, *National Electrical Code®*, 2014 Edition.
NFPA 780, *Standard for the Installation of Lightning Protection Systems*, 2014 Edition.
UL 497 series, *Protectors for Fire Alarm Signaling Circuits*.
UL 1283, *Electromagnetic Interference Filters*.
UL 1449, Third Edition, *Surge Protective Devices*, September 29, 2006.

A.2 NFPA Documents

1. *National Electrical Code® Committee Report on Proposals – 2013 Annual Revision Cycle*, National Fire Protection Association, 2012.
2. Marty Ahrens, *Lightning Fires and Lightning Strikes*, National Fire Protection Association, Fire Analysis and Research Division, June 2013.

© Fire Protection Research Foundation 2015   33
E. Davis et al., *Data Assessment for Electrical Surge Protective Devices*,
SpringerBriefs in Fire, DOI 10.1007/978-1-4939-2892-7

A.3 IEEE Documents

1. *How to Protect Your House and Its Contents from Lightning, IEEE Guide for Surge Protection of Equipment Connected to AC Power and Communication Circuits*, by Richard L. Cohen and others, ISBN 0-7381-4634-X, 2005.
2. Lightning and Insulator Subcommittee of the T&D Committee, *Parameters of Lightning Strokes: A Review*, IEEE Transactions on Power Delivery, Vol. 20, No. 1, January 2005.
3. Teru Miyazaki, et al, *A Field Study of Lightning Surges Propagating Into Residences*, IEEE Transactions on Electromagnetic Compatibility, Vol. 52, No. 4, November 2010.
4. Jinliang He, et al, *Evaluation of the Effective Protection Distance of Low-Voltage SPD to Equipment*, IEEE Transactions on Power Delivery, Vol. 20, No. 1, January 2005.
5. Shozo Sekioka, et al, *Simulation Model for Lightning Overvoltages in Residences Caused by Lightning Strike to the Ground*, IEEE Transactions on Power Delivery, Vol. 25, No. 2, January 2010.
6. Joseph Randolph, *Lightning Surge Damage to Ethernet and POTS Ports Connected to Inside Wiring*, IEEE, 2014.
7. Vladimir A. Rakov, *Direct Lightning Strikes to the Lightning Protective System of a Residential Building: Triggered-Lightning Experiments*, IEEE Transactions on Power Delivery, Vol. 17, No. 2, January 2002.

Note: The IEEE Power & Energy Society sponsors the Surge Protective Devices Committee, which provides information associated with their standards. Refer to http://pes-spdc.org.

A.4 NIST Documents

1. NIST Special Publication 960–6, *Surges Happen! How to Protect the Appliances in Your Home*, 2001.

Note: The NIST website provides many historical documents available in the public domain related to surge protection. Although this information is over 10 years old, it is still useful as a reference source. Refer to http://www.nist.gov/pml/div684/spd.cfm.

A.5 NEMA Documents

1. NEMA 2013 *U.S. Surge Protection Damage Survey*.
2. NEMA *Surge Damage Survey Results – Wave 2*, March 2014.

Note: The NEMA Surge Protection Institute maintains a website devoted to low voltage SPDs. Refer to http://www.nemasurge.org.

A.6 Insurance Industry Documents

1. *Lightning Sparks Concern For Insurance Industry; Homeowners Claims Rise Sharply Over Last Five Years*, Insurance Information Institute, March 2010.
2. *Thunderstruck! Average Lightning Claim Costs Up by 25 Percent, But Number of Claims Continues to Fall*, Insurance Information Institute, June 2013.
3. *Number, Cost of Homeowners Insurance Claims From Lightning Fell In 2013; Dry Conditions, Fewer Powerful Thunderstorms A Contributing Factor*, Insurance Information Institute, June 2014.
4. *Lightning*, Insurance Information Institute, August 2014.
5. *Protect Your Property From Power Surges*, State Farm website.
6. *Guidelines for Providing Surge Protection at Commercial Institutional and Industrial Facilities*, The Hartford Steam Boiler Inspection and Insurance Company.
7. *Approval Standard for Transient Voltage Surge Suppression Devices*, FM Approvals.

A.7 Manufacturer's Documents

1. Emerson Network Power Report SL-30119, *Surge Protection Reference Guide*, November 2011.

A.8 Miscellaneous Documents

1. A. Ametani,, et al, *Surge Voltages and Currents into a Customer due to Nearby Lightning*, International Conference on Power Systems Transients (IPST"07) June 2007.
2. Schneider Electric Data Bulletin DB03A, *Surge Protection: Measured Lightning Stroke Data*.
3. Al Martin, *Lightning Induced GPR, Why it's a problem, characteristics and simulation*, In Compliance, June 2012.

4. Thomas Key, et al, *Update on a Consumer-Oriented Guide for Surge Protection*, Proceedings, PQA'99 Conference, May 1999.
5. François D. Martzloff, et al, *The Role and Stress of Surge-Protective Devices in Sharing Lightning Current*, EMC Europe 2002, September 2002.
6. Arshad Mansoor, et al, *The Dilemma of Surge Protection vs. Overvoltage Scenarios: Implications for Low-Voltage Surge-Protective Devices*, Proceedings, 8th Annual Conference on Harmonics and Quality of Power, October 1998.
7. Air Force Manual 32-1181, *Design Standards for Interior Electrical Systems*.